GW00402456

MATHS TALK

Second Edition (National Curriculum)

THE MATHEMATICAL ASSOCIATION
and
STANLEY THORNES (PUBLISHERS) LTD

First published in 1987 by:
Stanley Thornes (Publishers) Ltd
Old Station Drive
Leckhampton
CHELTENHAM GL53 0DN
England

ISBN 0-85950-694-0

Reprinted 1988
Reprinted 1990
Second edition 1992

British Library Cataloguing in Publication Data

Maths talk. - 2nd ed
  1. Primary Schools. Curriculum. Subjects: Mathematics.
  Teachings
  I. Mathematical Association
  372.7044

  ISBN 0-7487-0557-0

Typeset by Tech Set, Gateshead, Tyne & Wear.
Printed and bound in Hong Kong.

Contents

A sub-committee of the Teaching Committee of the Mathematical Association was set up in 1981 to produce a booklet of guidance on Language in Primary Mathematics.

Its members were:

> Janet Duffin (chair)
> Ian Evans
> Linda Jones
> Zoe Rhydderch-Evans
> Hilary Shuard
> Sarah Somers
> Anne Wilson (secretary)
> Valerie Worcester

Members of the group have drawn on their own experience, and on the experience and expertise of others who freely gave their time and ideas to the whole group and to individual members of it. It is not possible to name everyone who contributed but they include:

> Tom Brissenden
> Dorothy Carter
> Roy Edwards
> Harriet Marland
> Julia Matthews

In addition, many people lent us tapes of children talking to each other and to their teachers.

We are especially grateful to the children of St Wulstan's RC First School, Stourport for their excellent drawings.

We thank Pauline Aston for typing and copying so many drafts of this book and for the cheerful way in which she did it.

We would like to thank the following for permission to reproduce copyright material:

The National Curriculum Council for the sections of non-statutory guidance for mathematics and English.
The Controller of Her Majesty's Stationery Office for extracts from *English in the National Curriculum, Mathematics in the National Curriculum* and *Aspects of Primary Education: The Teaching and Learning of Mathematics.*

Introduction

In this book we set out to show that the skills of spoken language are as important in mathematics as in other areas of the curriculum.

In the first edition we explored some recommendations of the Bullock Report and related them to the primary mathematics curriculum. In this new edition we follow up those ideas by considering recommendations in both the English and Mathematics components of the National Curriculum, and try to show common features of both where talking and listening can be seen to contribute to learning.

We asked ourselves a number of questions which helped to guide our thinking:

What part does discussion play in the development of mathematical thinking?

What are the benefits of talking and listening to one another?

How can mathematical teaching become more language-based?

What is the teacher's role in language-based mathematics?

What can children talk about that will help their mathematical development?

What changes in classroom management and organisation are needed for this style of working?

Can this style of teaching tell us any more about children's mathematical understanding?

Can we evaluate this style of working?

These questions are complex and answers are not easy to find, but we hope you will be interested in the suggestions we make.

The following quotations draw attention to the vital role which language plays in learning for both children and adults:

Often when we meet a problem we want to talk it over; the phrase 'talk it over' seems to imply something other than communicating ideas already formed. It is as if the talking enabled us to rearrange the problem so that we can look at it differently ... We cannot consider language in the classroom only in terms of communication, but must consider how children themselves use language in learning. The major means by which children in our

*schools formulate knowledge and relate it to their own purposes
and view of the world are speech and writing . . . The importance
of language – and of other symbolic systems such as mathematics
– is that it makes knowledge and thought processes readily
available to introspection and revision. If we know what we know,
then we can change it. Language is not the same as thought, but
it allows us to reflect upon thoughts . . . Thus children and adults
are not only receiving knowledge but remaking it for themselves.*

Douglas Barnes, *From Communication to Curriculum*
(Penguin, 1976)

*The use of appropriate mathematical language was also a key
factor in advancing the children's thinking and the quality of the
language used by the teacher in questioning and discussion was
of great significance. Where teachers were able to lead in this
way, purposeful talk about mathematical qualities was part of
the everyday language of the pupils, often matched by carefully
planned collaborative work, which helped pupils to extend and
develop their ideas about mathematics.*

*Aspects of Primary Education: The Teaching
and Learning of Mathematics* (HMSO, 1989)

1

Language: its impact on mathematics

> **Mathematics provides a way of viewing and making sense of the world. It is used to analyse and communicate information and ideas and to tackle a range of practical tasks and real-life problems.**
>
> *Non-statutory Guidance for Mathematics in the National Curriculum, A2*

Amongst adults it is accepted that the role of language is crucial both in the development of each person's thinking and in communicating the results of that thinking. An inability to communicate by talking, on the other hand, forms a barrier between people. Effective communication through spoken language goes far to promote confidence and the sharing and extension of tentative ideas. In daily life, discussion helps in the solving of problems.

In the pre-school years, children's informal learning and exploration of their environment is accompanied and illuminated by an unceasing flow of talk – sometimes monologue, sometimes questions and comments. Most teachers accept that, by striving throughout the primary years to enrich children's spoken language, they can make a valuable contribution to the development of children's thinking.

> **The skills of oral language are just as important in mathematics as they are in other aspects of the primary curriculum.**

Traditionally mathematics has been an exception to the use of children's talk as a vehicle for learning; mathematics lessons have not generally been associated with discussion or with communication between the learners through spoken language. Mathematics has a written symbolism of its own that can seem formidable. Perhaps this has caused some teachers to base their language work on the transmission and use of symbols, and on the learning of the formal spoken vocabulary related to the symbolism.

1

Instead of recognising that children need to make the spoken language of mathematics their own, teachers often provide a 'watered down' version of formal mathematical language, creating a 'patter' which they commonly use to help children to gain a tenuous hold on the symbols of mathematics and the operations these symbols represent. 'Patter' includes phrases such as '9 from 2 you can't', 'exchange a 10' and 'threes into 2 won't go, threes into 20'. These oft-repeated and unconsidered phrases may prevent children exploring the real meaning of mathematical language.

The National Curriculum in English and Mathematics

Just as the National Curriculum in mathematics is founded upon the Report *Mathematics Counts* (Cockcroft, 1982), so that for English is founded upon the Report *A Language for Life* (Bullock, 1975). These reports have had a remarkable influence over the development of children's learning in the years since they were published. We shall refer to the Bullock Report later.

The attainment target (AT1) in English, Speaking and Listening, can be matched to attainment target 1 (AT1) in mathematics, together with the corresponding programmes of study and non-statutory guidance.

At each level, the statements of attainment in English and mathematics which are particularly pertinent to developing maths talk are listed below. Appropriate examples drawn from mathematics are given for each group of statements of attainment. The close relationship between 'English talk' and maths talk is evident.

LEVEL 1 Pupils should be able to:

English
- participate as speakers and listeners in group activities, including imaginative play.
- listen attentively, and respond, to stories and poems.

Mathematics
- talk about their own work and respond to questions.

Children try to find out who in their group is tallest, who is next, and so on down to the shortest. They tell each other how they know, and discuss how to record their findings.

Act out and sing *Ten green bottles hanging on the wall.*

LEVEL 2 Pupils should be able to:

English
- participate as speakers and listeners in a group engaged in a given task.
- describe an event, real or imagined, to the teacher or another pupil.
- talk with the teacher, listen, and ask and answer questions.

Mathematics
- talk about work or ask questions using appropriate mathematical language.
- respond appropriately to the question: 'What would happen if . . . ?'

Tell a friend in what order you put your clothes on this morning. Make some patterns in colour, using linking cubes. Explain to others in the group what you have done, and ask others about their patterns. Imagine what a different pattern would look like and tell us about it.

LEVEL 3 Pupils should be able to:

English
- convey accurately a simple message.
- listen with an increased span of concentration to other children and adults, asking and responding to questions and commenting on what has been said.

Mathematics
- use or interpret appropriate mathematical terms in a precise way.

Explain to a visitor how to go to Mrs Smith's class.

Two children went to an adventure park for the day. Pupils in their group listen to their account of what happened and question them about the events.

What are the chances of rain when we go on our trip to the seaside on Thursday?

LEVEL 4 Pupils should be able to:

English
- give a detailed oral account of an event, or something that has been learned in the classroom, or explain with reasons why a particular course of action has been taken.
- ask and respond to questions in a range of situations with increased confidence.

- take part as speakers and listeners in a group discussion or activity, expressing a personal view and commenting constructively on what is being discussed or experienced.
- identify and obtain information necessary to solve problems.
- participate in a presentation.

Mathematics
- give some justification for their solutions to problems.

Explain a way of adding, say, 19 to various numbers, without writing anything down.

Visit the town museum and then report back to the rest of the class, describing the places mentioned, the activities available, the entry cost and the times when the museum is open.

Plan how to find out what happens to the shadow of the flagpole in the playground during the day, and then carry out the plan.

Four children are asked to go into the playground to observe the passing traffic for half an hour and then to make a presentation to the class about what they found out.

LEVELS 5 and 6 Pupils should be able to:

English
- give a well organised and sustained account of an event, a personal experience or an activity.
- contribute to and respond constructively in discussion, including the development of ideas; advocate and justify a point of view.
- use language to convey information and ideas effectively in a straightforward situation.
- contribute to the planning of, and participate in, a group presentation.

Mathematics
- interpret information presented in a variety of mathematical forms.
- pose their own questions or design a task in a given context.

Every child plants a bulb one day in October. Discuss and plan how to record the growth of the bulbs.

In a series of numbers 1, 2, 4, ... a child says what he or she thinks is the next number, and why.

Two pupils explain to a group how they worked out the number of tiles needed to resurface the classroom floor.

The class plan and present an assembly on shape.

Programmes of study for Key Stage 1 in English can be applied throughout Key Stages 1 and 2 in mathematics.

English KS1   Through the programme of study, pupils should encounter a range of situations, audiences and activities which are designed to develop their competence, precision and confidence in speaking and listening . . .

In mathematics we can provide situations, audiences and activities where children talk about their activities and the materials they are using in pairs, and in groups, to the whole class, to their teachers and to visitors.

We can develop:

competence   in organising activities and solving problems;

precision   by careful unambiguous use of language and accurate number work;

confidence   by encouraging tentative answers, with opportunities for a change of mind, so that pupils feel they have been successful.

The orders and recommendations made under the Educational Reform Act can be seen to reinforce the advice given in the Bullock Report, which is still as valid as ever.

The Bullock Report: its impact on primary mathematics

The Bullock Report (in para. 5.30) lists eleven specific uses of language which young children need to experience, and suggests that these form a basis on which the teacher can monitor children's progress in language. Language-based mathematics teaching should give experience of each of the items on the Bullock list. Although the Report put forward this list only in its discussion of the language of young children, many items are also appropriate in the language needed by older children in talking about mathematics. The Bullock list, with examples from mathematics, is given below.

1   Reporting on present and recalled experiences:

A group of young children recall what happened yesterday when they attempted to partition 6 in as many different ways as possible using Cuisenaire rods. They compare this with today's task in which they are using Unifix cubes to achieve the same objective.

An older group recall and report on their tallying of yesterday's traffic, in preparation for drawing a graph of the information.

2 Collaborating towards agreed ends:

A group of young children plan together how they will investigate which numbers are exactly divisible by 3. They decide together what needs to be done and allocate tasks to each member of the group.

A group of older children share out the writing-up of their investigation of the number of squares through which diagonals of rectangles pass when they are drawn on square grid paper. Each contributes an account of a different example, one tackling squares, another even-by-even rectangles, and a third odd-by-even.

3 Projecting into the future; anticipating and predicting:

A group of young children examine the number of children on roll in the school. They look at the admission book of the school and try to anticipate what size the school will be in three years' time. An older group tackle the same problem by making enquiries in the infants' school and investigating movement of families in and out of the neighbourhood.

4 Projecting and comparing possible alternatives:

A group of children try to improve the organisation of the dinner sittings in the school. Some children have school dinners and others bring a packed lunch. The canteen has a restricted number of seats, and the numbers of children having dinners and packed lunches fluctuate.

A group of children investigate routes from their classroom to the headteacher's room.

5 Perceiving causal and dependent relationships:

Young children group cubes in twos, and find that alternate numbers of cubes can be grouped without remainder. The words 'even' and 'odd' are introduced. They do the same activity using groupings in threes.

Older children explain why the squares of even numbers are always multiples of 4.

6 Giving explanations of how and why things happen:

Young children put a heavy object on one side of a pair of scales and a light one on the other – one side goes up and one goes down. They explain why this happens.

An older child explains how and why the digits move when a number is multiplied by 10: 'each one becomes a ten; and each ten becomes a hundred'.

when we do Weighing the heavy things go down and the light things go up

7 Expressing and recognising tentativeness:

A young child guesses which of two containers will hold more water.

It is hard to tell wich holds the most. I think that the mug might be the same as the tall bottle.

An older child strives to discover the factors that might affect the period of swing of a pendulum, saying 'It might be the length of the string'. The tentative hypothesis is then checked by experiment.

8 Dealing with problems in the imagination and seeing possible solutions:

A child builds a 7 × 7 × 7 cube, works out that there are 49 units in a layer and sees that this is one less than 50. This enables the total number of units in the cube to be seen as 7 fifties less 7.

9 Creating experiences through the use of imagination:

On the table there are 6 blue cubes and 4 red cubes. After combining the two sets and finding that there are 10 cubes altogether, a group of children make up number stories about this. The stories are of the type: 'I went to the shop and bought some bubble gum for 4 pence. I got 6 pence change from a 10 pence piece.'

Older children plan a class outing, using maps and timetables.

10 Justifying behaviour:

(In mathematics, this could be seen as justifying the correctness of an action or process.)

A child has tackled the task of finding the possible ways of arranging three identical cardboard squares in different patterns so that they touch one another on at least one side. Having completed the task, the child justifies the solution to the group, proving that there are no other possibilities.

An older child does the same activity with four or more squares.

Ways of arranging 6 squares

11 Reflecting on feelings, their own and other people's:

A problem solving and investigational approach to mathematics causes children to experience both some pleasure in success, and some frustration at temporary failure. The experience of sharing these feelings with teacher and peers helps children to realise that both initial failure and eventual success are necessary to intellectual growth.

Mathematical language at home

Mathematical language may play little part in the conversation of the home, although the presence of calculators and computers in many homes is slowly beginning to have an effect. Many children are disadvantaged linguistically in mathematics in comparison with other areas of their experience. Most children, therefore, need particular help with language in mathematics.

Children's early mathematical language and thinking

Mathematical thinking grows with the natural language which children develop in the pre-school years. Some of this language has considerable mathematical potential and significance. Aspects of early speech which are important to mathematical development include:

- **attribute** words such as 'big', 'little', 'blue';
- **position** words such as 'on', 'in', 'under', 'behind', 'from', 'to';
- **comparison** words such as 'bigger', 'more';
- **question** words such as 'where', 'who', 'when', 'why';
- **connectives** such as 'but', 'because', 'and'.

Young children find some of these words more difficult than others. For example, 'big' eventually has to be seen as a relative term, and is often replaced by 'bigger than' or refined as 'taller than' or 'wider than' . . .

The development of these ideas and their associated vocabulary is vital to children's mathematical development. Mathematical thinking rests on these starting points.

MATHEMATICS FROM CHILDREN'S EXPERIENCE

Most early speech comes from experience gained in play with other children and through interaction with adults.

**Mathematics is best developed through activities
which use the language of mathematics
and build up mathematical thinking.**

It is now recognised that mathematics is best developed through activities which use the language of mathematics and build up mathematical thinking. However, this principle is often neglected in mathematical work given to older primary children; there is evidence that even adults perform better in mathematics if it is related to their work or interests. Activities that are relevant to children's lives are more likely to generate mathematical talk and thinking than teaching which concentrates only on mathematical operations and symbols.

EXPERIMENTING WITH WORDS

Another feature of young children's experimentation with language can help teachers to provide a fruitful learning environment even for older children. Young children constantly try out new words and phrases, either by imitation of others or by experimentation after listening and reflecting. Unlike adults placed in mathematical situations, children rarely exhibit fear of being wrong in their experimental use of language. It is important to encourage a similar attitude in the classroom. Mistakes need to be seen by teacher and children alike as learning agents to be accepted and built upon, not as evidence of failure. When discussion becomes an integral part of the work in mathematics, unexpected remarks often generate new ideas to be welcomed and encouraged:

- Some words have a special meaning in mathematics which is different from the meaning in everyday usage, e.g. face, table, volume;

- some words have more than one meaning within mathematics, e.g. base, difference, light, square.

Sometimes a child learns to use a word without fully understanding its meaning, and the teacher is misled into believing that the concept to which the word relates is understood. Attention to language skills in mathematics alerts the teacher to this problem, and helps to overcome it.

Modifying teaching styles

Traditional teaching methods need to be modified to include more discussion. A common sequence in classroom interaction is:

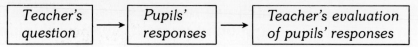

| Teacher's question | → | Pupils' responses | → | Teacher's evaluation of pupils' responses |

The teacher has a plan for proceeding through the lesson, and evaluates the children's responses according to their closeness to a 'correct' answer. Pupils often feel afraid of making mistakes, thus limiting the kind of classroom discussion that will aid understanding.

In looking for ways of modifying classroom interaction, the teacher has to take on a new and unfamiliar role: that of participant rather than of leader and evaluator. The children will also need to adapt to the new style. Of course, the teacher will sometimes need to use the more familiar mode of 'exposition'; 'discussion' is an additional teaching style, and the children need to distinguish between these forms of interaction. In the infant classroom where group arrangements are varied for different types of lesson, the children understand that the mathematics talk now includes their contributions, as does the 'news' period; in classes of older children it may not be so simple. It may be necessary to have a different class grouping for discussion, or the teacher might indicate the change of style by sitting instead of standing, or moving to another position in the classroom.

A new model of learning style might be:

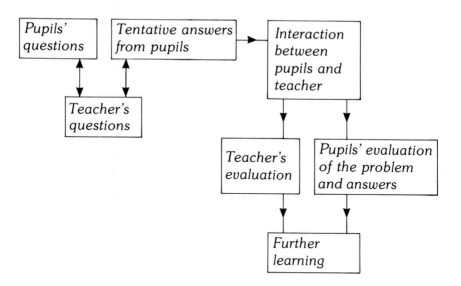

NEW SKILLS FOR TEACHERS AND CHILDREN

An inexperienced teacher may inhibit children's talk by moving on as soon as the 'right' answer has been given. Both teacher and children benefit from listening to and giving other answers and explanations. Children gain confidence and also acquire new skills of sharing and helping, explaining and listening, and of co-operating instead of working in isolation. Such skills have to be learned. Teachers must encourage children to talk, and learn to listen carefully to them.

| Listen to the children. |

INFLUENCE OF NEW TECHNOLOGY

Calculators and computers do much to promote talk among small groups of children. The teacher can join in the discussion, allowing the children to explain their ideas, which helps them to clarify their thoughts. The introduction of problem solving and investigations can also be exploited more effectively if discussion skills are already well developed.

ASSESSMENT THROUGH CLASSROOM TALK

In addition to providing enhanced opportunities for children to learn mathematical ideas, another valuable outcome of a discussion-based teaching style is that the necessary assessment can be achieved without threat. As teachers and children become more proficient in the new skills, classroom talk can show the teacher how children's mathematical thinking is developing, besides revealing misunderstandings and why they occur.

The following chapters aim to illustrate the principles outlined here, and suggest some solutions to the problems we have identified.

2 *Listening to children talking*

Transcribing taped discussion is very time consuming, but transcripts of classroom talk can be a fruitful source of information about the nature and quality of the discussion teachers have with their children. Many such transcripts are available for examination as more and more teachers and educators become aware of the importance of talk in learning mathematics.

Transcripts can provide information which a classroom observer might otherwise miss: about the words being used, about the discussion model the teacher is employing and about other characteristics of speech which might otherwise be unobserved in a busy classroom. On the other hand, a transcript cannot tell us about the tones of voice, the gestures, the facial expressions of the children, or about their relationships with each other and with their teacher.

The transcripts discussed below demonstrate a number of the communication issues outlined in Chapter 1.

Apparatus with a four-year-old – only the teacher talking

Naveed – aged four – is working with his teacher using pieces of Unifix connected into lengths of 1, 2, 3, 4, 5 units.

T *Right, this is Naveed and Naveed is four, aren't you?*

Here we go: 1, 2, 3, 4 (handling the equipment, child copying).

Good boy. And we're going to see if Naveed can match some things. Now look, see what we've got here. Can you find one like it? (Teacher holds up one cube.)

Where's your one? Good boy. That's one.

C *One.*

T *Right, now we go on to two. And we've got two: 1, 2. Where is it, Naveed? Have you got two? Which one? One like that? Good. Look, that's the same as two. Put that one there. Now look we've got three. 1, 2, ... Where's one like that? Which one's the same?*

Notice that here all the speech, except for one word, comes from the teacher. At first sight it is tempting to think that Naveed is at the stage when he understands spoken language but is not yet ready to speak. However, this could be too simple an explanation: perhaps the teacher's actions, gestures and words render speech unnecessary for the child, or perhaps the child is intimidated by an unfamiliar situation. The impression is given that the apparatus was already prepared so that the child could not become actively involved.

Perhaps, then, the use of toys in a more informal setting might have been more effective in encouraging Naveed to talk.

The teacher's speech, as revealed by the transcript, indicates a possible source of confusion for the child in spite of the simple language being used: the word 'one' is used both for the single cube and for the 2-cube and 3-cube lengths. Though Naveed gives no indication of being confused, it is important to be aware of the danger. Thus the transcript has helped to expose a potential confusion.

Coloured rods for number difference – conflict of meaning

The children in this example are older and more articulate. The teacher is using coloured rods to elucidate the 'difference between' two numbers by comparing the rod lengths.

A 5-rod and a 9-rod have been compared.

T   *So what's the difference between these two? What's the difference in number? . . . The difference is . . .*

C   *9 and 5.*

T   *And what's the difference between 9 and 5?*

C   *9 is longer.*

T   *Yes, 9 is longer and 5 is . . .*

C   *Lesser.*

T   *Less than 9 is it, so what's the difference between them?*

You told me before that the difference between the other two was 2: so what's the difference between these two?

C   *4.*

T   *It's 4 isn't it? Right – so what's the difference between these two?*

C   *The red one is bigger than the yellow one.*

Clearly the teacher is intent on numerical difference while the children seem to find it easier to focus on the non-numerical differences in length. Notice that the teacher modifies the question in order to focus attention on difference in number, indicating a recognition of the children's different perception. But, in spite of this, the children fail to recognise what the teacher seeks and they persist with their own non-numerical comparison. By reminding the children of an earlier example the teacher then manages to establish a numerical difference, only to find that they revert to their earlier length comparison.

This transcript shows the importance of using appropriate materials for the concept in question. In this case the materials used generate a mismatch between the perceptions of teacher and children which the ensuing discussion does not resolve.

Both teacher and children show persistence in retaining their own, conflicting, perceptions of the material used. Both perceptions are valid but for different mathematical purposes. The teacher might have been more successful in developing the numerical difference through countable objects, instead of rods which may be more appropriate to concepts of measurement.

Mismatch of perception can be valuable in mathematical development when it is recognised and exploited. If it merely forms a source of misunderstanding between children and teacher it is counter-productive.

Cueing as a teacher/pupil convention

Notice, too, that the children's answers are monosyllabic, the teacher providing most of the speech. 'Cueing' occurs when the teacher pauses while waiting for the children to fill in the missing word. Cueing, which recurs in many of the transcripts examined, seems to include conventions which are clearly understood and used by both children and teachers.

Cueing with reception children

Some reception children have been discussing with their teacher the migration of swallows to Africa and are planning to produce a frieze of birds waiting to set off on their flight. Mathematical shapes are to be used to make models of the birds. Here mathematics comes from a real situation and is integrated into more general learning.

T   *It's a rectangle – so you've got two very long . . .*

C   *Rectangles.*

T   *And thin and . . .*

C   *Narrow.*

T   *A forked tail, so it means . . . some birds' beaks are facing . . .*

C   *Ahead.*

T   *Ahead, yes . . . and you've printed this one looking down into the ground. It looks as if he's looking . . .*

C   *Down.*

T   *Down to the ground. This one is looking . . .*

C   *Up.*

Cueing here fails to encourage the children to extend their talk beyond the monosyllable required to complete the teacher's sentence, and the opportunity created by the imaginative setting and worthwhile activity is lost. In the following example both children and teacher use a cueing technique and both are frustrated by the experience.

Talking about 'paper and pencil' sums – problems from patter

The teacher is going through some sums with a group of children from a Year 4 class.

T   *Do you remember when we went through these sums last week? What we said we had to do if you hadn't enough?*

C   *Er.*

T   *What did you say we had to do?*

C   *You take the . . . er . . . that . . . you had to borrow . . .*

T   *What did you have to borrow?*

C   *3.*

T   *Did you borrow 3? You had 4 tens and 2 ones. You've got all that written down. What did you have to borrow?*

C   *4 . . . 1 . . .*

T   *Well, which number did you need to make bigger when you were taking the 5 away? Did you need to make the 2 ones bigger or the 4 tens bigger? What did you need to do?*

C *Make the 2 bigger?*

T *You had to make the 2 bigger – that's right. How could you make the 2 bigger? . . .*
   Well, you told me last week when you were doing it. You had to borrow . . . one of these tens. Yes, you could do it on paper. Isn't it funny – suddenly you can't do it?

Confusions exist in this transcript at several different levels; perhaps the first one is that of the reader who does not know the exact 'sum' under discussion. From the children's initial response of '3' as the number to be borrowed, together with the teacher's reference to '4 tens and 2 ones' and to taking away 5, it seems likely that the calculation involved is:

$$\begin{array}{r} 42 \\ -25 \\ \hline \end{array}$$

Now the child's response of '3' does not seem quite so nonsensical for it could indicate that in trying to take 5 from 2 you are short of 3. But that useful perception does not fit normal paper and pencil subtraction so it is discarded by the teacher, and, because of the confines of the exercise, it cannot be pursued with advantage.

The teacher is trying to get the children to explain a process they can apparently do on paper. The children are trying to reproduce the procedures they have learned, using the patter word 'borrow'. This word, still widely used, has no association with the subtraction, as can be seen when it is related to concrete apparatus.

Notice that the children, too, use cueing to encourage their teacher to give them more clues; they sometimes guess when they are pressed to respond.

Undoubtedly cueing is well-intentioned on the part of both teacher and children, but often it inhibits further response from the children. To the children it can seem like a 'guess what I'm thinking' exercise and, anxious to please, they try to supply what is required.

We learn from this transcript that moving too soon to formal recorded calculations causes difficulties which can be compounded by adhering to artificially created patter.

Examples of more effective cueing devices

Some seven-year-old children were looking with their teachers at different types of plastic tape used for binding parcels. They were focusing on the differences between the tapes.

T   *What differences are there?*

C   *One is blue.*

T   *Yes.*

C   *One is white.*

T   *Yes.*

C   *One is blue and one is white.*

T   *Yes, anything else?*

C   *The blue one is bigger?*

T   *How bigger?*

C   *Longer.*

T   *What is longer than what?*

C   *The blue one is longer than the white one.*

T   *What about the white one?*

C   *The white one is shorter than the blue.*

Notice again that while the children are perceiving differences which are not numerical, they are encouraged by using the cueing effect of 'yes' to pursue the original line of thought, and, eventually, to articulate the differences adequately and correctly. In this case the teacher's style of cueing is enabling rather than inhibiting, with the children saying more than the teacher.

Unobtrusive introduction of mathematical vocabulary

The 'swallows' lesson also contained passages with more effective cueing in them.

T   *Right, a sponge in the shape of a cylinder. Did you use the whole of the sponge to print it?*

C   *No – the face.*

Notice how the child, in introducing the mathematical term 'face' referring to a solid shape, appears to have overcome possible confusion with its everyday meaning.

T   *Which face did you use?*

C   *The round one.*

T   *The round one . . . the circular face . . . good. Now . . . what did we use to print the head with?*

This time the teacher unobtrusively introduces another mathematical term without interrupting the flow of the conversation, and at the same time manages to encourage the child.

Later the term is again used by the teacher.

T   *Which face?*

C   *The one with holes . . . the one with the circle.*

T   *The circular face.*

Now the teacher reinforces, by using again the mathematical term which the child cannot yet produce spontaneously, although the use of the word 'circle' somewhat hesitantly suggests that the word 'circular' is nearly within the child's grasp. The opportunity given to children by this kind of interaction to try out the use of new words in an encouraging environment seems to be a better way to achieve successful learning of new vocabulary than having its meaning firmly given, prior to seeing it written up on the board to be copied and memorised. This way children are much more likely to begin to introduce new words into their speech when they feel ready to do so, and, moreover, to have an accurate perception of their meaning.

Spontaneous language from an activity

I wrote my name on a piece of paper and when I put it to the mirror it went backwards.

in the mirror

I put the mirror like this   I Saw three girls altogether

When children are encouraged to speak naturally about what they do, their speech can show delightful signs of language and mathematical growth. In the next transcript a very young child is using a mirror with a book*. The teacher turns the pages when there is a pause; the child does all the talking.

That's a whole girl, when we put it there, it's a whole boy. Can you see? Look – a whole boy, can you see? Can you see?

That's a whole boy.

That's a whole girl.

That side she's happy and that side she's happy.

That side's the same as that one on the mirror and that side's the same as that one on the mirror.

On that side they're the same socks.

The cat is the same when you shine the mirror, it's the same cat.

It's the same teddy back together when you put it like this.

Nobody is on that swing and there is someone on that swing.

When you do it like this there's two girls – twins!

When you put it on this side there's two and when you put it on this side there's two and when you put it on this side there's three.

*Walter, *A Second Magic Mirror Book* (Scholastic Publications, 1984)

These spontaneous remarks arose from something which had caught the child's interest at that moment. They express ideas and use language which will be valuable in later years.

Children reporting on an activity

Some older children, in Year 4, are talking with their teacher about some work they have been doing. They have been comparing the capacities of different containers and are telling their teacher what they have found out from their activities.

C *Well, this one was 23.*

T *23 what?*

C *23 matchboxes – but we thought this one would be bigger than that one. This one was 26. So it's 3 more matchboxes.*

T *Why did you think this one was going to be smaller?*

C *Because it was square. It's quite fat but I thought they would hold sort of like the same 'cos it's thinner and this is just a bit taller.*

T *But they're very close together. 3 matchboxes of sand isn't very much sand. What else did you notice about these – this tall flour container?*

C *Well I reckoned that it was 20 but it's 18.*

T *Hmm. In comparison to this one perhaps. How much was that one?*

C *One more.*

T *And so they were very close. Did you have a job to decide which was the bigger of those two? No, you were quite sure? Because a lot of children, because this one is very tall, put that one before that one – a lot of the other children have done that.*

C *That's thinner than that.*

T *You're obviously on to that, aren't you?*

C *Before we done it there was a slight problem. We were measuring it and the sand was coming out of the bottom and we had to tip it all out and Sellotape it all up.*

Now the teacher is seeking information from the children rather than asking questions to which she already knows the answers. Immediately the discussion changes its tone: the children are the talkers, the explainers, and their confidence also allows them to speak of a difficulty they have encountered. They are clearly experiencing a different quality of discussion, and it begins to approach more nearly the type of discussion which, in ordinary life, leads to learning.

> **The teacher's job is to organise and provide the sorts of experiences which enable pupils to construct and develop their own understanding of mathematics, rather than simply communicate the ways in which they themselves understand the subject.**
>
> *Non-statutory Guidance for Mathematics in the National Curriculum, C2*

Teachers' Changed Roles

> **In covering the programme of study for Speaking and Listening, the teacher should be:**
>
> - **helping to sustain what children are trying to say by showing interest;**
> - **an exploratory user of language;**
> - **supportive and encouraging to the children in their use of language;**
> - **able to create an atmosphere of challenge and involvement;**
> - **prepared to intervene only when it is appropriate;**
> - **aware of the special needs of speaking impaired and/or hearing impaired children;**
> - **sensitive to individual needs, especially when the child is shy or lacking confidence;**
> - **aware of the need to make the verbally aggressive or dominating child sensitive to others;**
> - **aware of the need for bilingual children to work with others in their home language and in English, to strengthen their capacity to use English for a range of purposes;**
> - **prepared to monitor and evaluate the children's use of spoken language.**
>
> *Non-statutory Guidance for English in the National Curriculum, C8*

Many teachers may need to adapt the teaching styles they already use in teaching English, to foster discussion in mathematics. The Bullock Report reinforces the guidance offered in the National Curriculum:

- The need to negotiate meaning through classroom discussion (general theme of Chapter 10 of the Report).

- Much classroom talk consists of a three-stage process: teacher question, pupil response, teacher evaluation (para. 10.3).

- Teacher evaluation of their contributions has an inhibiting effect on many pupils (para. 10.4).

- Classroom 'discussion' is often no more than the pupil trying to read the teacher's mind (para. 10.3).

- The teacher needs to take up and use the pupil's contribution (para. 10.2).

- The teacher needs to be a good listener (para. 10.11).

- The teacher's receptive silence is often as important as utterance in stimulating pupil talk (para. 10.11).

- The pupil's own language must be accepted (para. 10.5).

- The pupil's language indicates where he/she is in progression from the familiar to the new (para. 10.2).

- The teacher needs to devise situations and structure learning so that the child becomes positively aware of the need for a complete utterance (para. 10.11).

- Small group talk should have as its end sharing and communicating with the whole class (para. 10.12).

- The tape recorder is an indispensable instrument for oral work, and no teacher should be without ready access to one (para. 10.23).

In this chapter we have described how recording children talking can help us to analyse some of the problems of communication which can occur during classroom talk, and we have given some examples of profitable discussion. In the following chapters, ways will be suggested for restructuring mathematics lessons in order to enhance opportunities for language development within and through mathematics.

3 Experience, language and learning

> **Activities should be balanced between different modes of learning: doing, observing, talking and listening, discussing with other pupils, reflecting, drafting, reading and writing, etc.**
>
> *Non-statutory Guidance for Mathematics in the National Curriculum, B9*

Introduction

The lessons in which the teacher is the 'talker', while the children are silent or monosyllabic, are not suitable for the development of mathematical language. More fruitful learning takes place when the pupils talk to one another and to the teacher in order to explore and to learn mathematics, with the teacher listening to the children, thereby learning more about each child's thinking. In this way the teacher can make professional decisions about appropriate activities to help develop children's thinking. Sometimes a child's thinking may be mistaken, sometimes it may be a legitimate alternative to the teacher's thinking, and sometimes the teacher may be surprised by its insight and penetration. Greater understanding of children's learning can be gained by listening to them and encouraging them to ask questions. It is not always easy for teachers to move from a situation in which they use carefully pre-planned ideas and techniques. It is not immediately obvious how to reorganise the classroom so that the children can talk freely about mathematics. The following points may help to smooth the transition.

Generating talk from experience

From a very early age some of the main sources of language exchange are children's activities, the things that have happened and are happening to them and to those around them; it is experience which generates language exchange and communication.

Children's learning through experience can be used by the teacher to generate 'mathematics talk' and to help children develop mathematically. The natural form of learning can be expressed in the following model:

EXPERIENCE → LANGUAGE → LEARNING

Children gain experience through seeing, hearing, touching and handling; all these can be the basis for classroom activity, talk and learning in mathematics.

Moving house

Which is the heaviest thing to carry?

Paragraph 306 of the Cockcroft Report states that:

> There is a need for more talking time; ideas and findings are passed on through language and developed through discussion, for it is this discussion after the activity that finally sees the point home.

Cockcroft, *Mathematics Counts* (HMSO, 1982)

We believe that class talk is as important during the activity as it is afterwards.

WHAT SHALL WE TALK ABOUT?

Mathematical talk can be generated from a wide variety of activities and experiences. Here are a few suggestions:

a range of water containers,
a set of three or four floor tiles,
pieces of wood: oak, pine, balsa,
a number line,
a die or several dice,
a six-pointed star,
a house brick and a polystyrene brick.

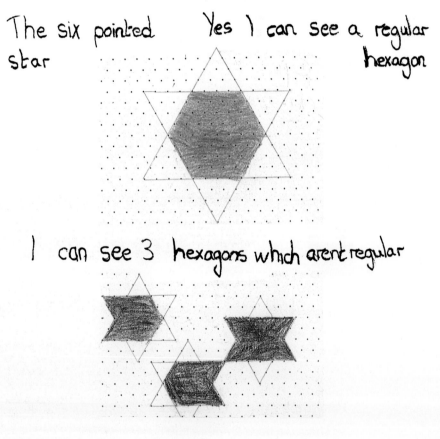

The six pointed star

Yes I can see a regular hexagon

I can see 3 hexagons which aren't regular

An overhead projector may help to demonstrate to the whole class a particular idea discussed by one group. It is important for the interests of the children to be captured by a problem, perhaps one which they have generated themselves. It is not always necessary for the teacher either to know the direction which the investigation will follow or its solution. Children talk more freely when they are involved in an activity about which they have become excited.

Which covers the largest surface

8 tiles can cover a surface in several ways.

How many times can I throw a six?

Even watching other children experimenting can be helpful; many children seem very keen to explain their activity to others in the group.

Types of 'mathematics talk'

1 Teacher-led

This method is particularly appropriate to the introduction of new ideas and skills, the teacher's explanation often helping children to focus on the major points.

However, children may search for responses which please the teacher, rather than focus on the thread of the argument; responses of this type can be misleading because children seem to have understood when they have not. Children's language will not develop if they only give short answers to closed questions.

2 Child-teacher

This is the most difficult style of classroom interaction for the teacher to manage if genuine discussion is to take place. Such discussion can only be sustained if the children have sufficient confidence in themselves and are sure that their contributions will be taken seriously.

> **Avoid telling children they are wrong.**

It is important for the teacher to avoid telling children that they are wrong – this immediately changes the conversation into exposition. In discussion, when children realise that other people hold different views, they have the opportunity to reassess their own thinking or to argue their own case.

> **Pupils should be encouraged to undertake work in teams, discussing mathematical problems, evaluating ideas and alternative solutions, and jointly finding ways to 'crack' an onerous assignment.**
>
> *Non-statutory Guidance for Mathematics*
> *in the National Curriculum, B9*

3 Discussion between children

This often takes place vigorously in the absence of the teacher, but it may at first be difficult to sustain when the teacher joins the group; many children immediately appeal to the teacher for approval and leadership. If the children are sufficiently interested in the problem they are tackling, and if they have become accustomed to mathematics talk which the teacher does not always lead, then the skilful teacher may be able to introduce further points and stimulate continued thought without dominating the proceedings. Many teachers have also found that the computer is a stimulus to vigorous discussion amongst children. If small groups of children work together at a suitable problem-solving program, mathematical discussion and argument will be stimulated.

Accepting every child's contribution

For a true exchange between teacher and children, the teacher must create an encouraging and receptive classroom climate. The children need to know that they are valued members of the group, and that their contributions are taken into account.

> **The teacher must create an encouraging and receptive classroom climate.**

In addition, the work needs to be carefully structured so that the children grow in confidence through contributing their own ideas to discussion, until they are mature enough to accept that they may sometimes be wrong, which is a valuable part of learning.

Barriers are often erected between teacher and children by the premature use of technically correct language. Some children quickly absorb the formal language of mathematics, but its use does not always guarantee that it is understood; too early an insistence on 'correct' usage may also make children unwilling to use their own language to express their ideas.

As Barnes says:

> *The attempt to predetermine the terms which the children use may prevent them from using language to learn with: the control affects the learning.*
>
> Barnes, D., *From Communication to Curriculum* (Penguin, 1976)

In order to negotiate meaning children need to experiment with language, trying it out until their own meanings match those of their peers and their teacher.

Here is an example of the way in which children's language might develop from the spontaneous speech arising from activity to a more mathematical form:

> A group of children are sitting around a model of an irregularly shaped lake which is much narrower at one end. They are discussing where it would be best to try to jump across.
>
> 'I think it would be best to jump across here where it is thin', says a child pointing to the narrowest part of the lake.
>
> The teacher accepts the word 'thin', saying 'Yes, that's right, where it is thin, where it is narrow'.
>
> On subsequent days the children are given the opportunity to compare and contrast ribbons of varying widths, knitting needles of different gauges, lines made with paint-brushes of various sizes, different widths of streets and lanes, etc.
>
> As the children use the words 'wide', 'narrow', 'fat', 'thick', 'thin', and their comparative forms, understanding, and the language needed to express that understanding, will gradually become more precise.

In all three types of 'mathematics talk' (page 28), one of the most difficult things of all for both teacher and children is learning to accept every child's offerings. Children tend to glance at the teacher in order to judge the response both to their own offering and to that of other children. It is not until they can talk to each other in the teacher's presence without seeking that small reassurance, and yet, at the same time, accept the teacher's offerings alongside each other's, that the transition from 'the teacher knows the right answer' to 'we are all finding out together' will be achieved. Some suggestions are made in the following pages.

GIRLS AND MATHEMATICS

Recent reports suggest that girls may respond especially well to discussion. Cockcroft referred to research which suggests that:

> *Girls may have a greater need than boys to develop understanding through discussion, and many girls consider that mathematics teachers do not listen sufficiently to what girls say in response to questions.*

Whilst The Royal Society Report *Girls in Mathematics* (1986) recommends that teachers:

> *Encourage all pupils to talk about mathematics and attempt to bring a "social" (perceived as "female") dimension to the teaching. Introduce group work and co-operative teaching styles; do not allow boys to dominate mixed groups or girls to defer to boys in discussion.*

CLASSROOM ORGANISATION WITH YOUNG CHILDREN

Some changes in classroom organisation may be necessary to enable talk to be used in developing mathematical understanding. For a teacher of younger children, this may mean a fairly simple adaptation of the methods used for other aspects of language development. Children and teacher gather together, and the teacher introduces a stimulus for mathematical discussion, or asks the children to recall, and talk about, what they have been doing. Similar discussions can take place when the teacher joins a group of children who are engaged in a mathematical activity or game.

CLASSROOM ORGANISATION WITH OLDER CHILDREN

For a teacher of older children, however, a more radical re-thinking of classroom organisation may be needed to provide opportunities for 'mathematics talk'. Children will need to be organised so that they can work and talk together in groups for some parts of the mathematics lesson. This idea of group work is not new, but the phrase 'group work' often describes the way that children are seated in the classroom; groups are seated together at tables, but children work individually at their own task. Arthur Owen, HMI, coined the phrase 'the loneliness of the long-distance workcard' to describe this type of 'group work'. Children who work individually in this way have little encouragement to discuss the mathematics they are doing, and the teacher of a class of thirty has little time to join in mathematics talk with each child. This kind of 'group work' is unlikely to achieve much development of children's mathematical language.

However, co-operative group work and 'mathematics talk' can be introduced into older children's mathematics in several ways. For example, the teacher may first set the scene with the whole class, describing the tasks or investigations to be tackled and posing questions for the class to think about; these tasks should be designed to encourage the sharing of the mathematical ideas inherent in the activity. The class can then be set to work in groups, each group pursuing its own ideas, while the teacher circulates and observes.

A GROUP WORKING TOGETHER

Groups of between three and six are usually most successful. Children in a group must work together at a single task if they are to discuss their ideas. There is sometimes scope for setting up groups of mixed attainment, rather than always using homogeneous groups. It is worthwhile trying to work out the expected progress of the various activities, and trying to visit each group at a crucial stage. The teacher will also have to encourage the group members to talk together – however surprising this may seem – and children may need to be taught how to do this.

When the teacher notices that at least one group is beginning to reach some conclusions, a further whole-class session can be a focus for the sharing of ideas, with contributions from each group in turn. In this type of organisation, it is important for the teacher to ensure that the key ideas which emerge are shared and discussed, because some children may have yet to develop them.

In planning, the teacher can divide the class into groups, which should be so arranged that each child can see the focal point of the discussion, be it apparatus, an OHP or a blackboard. A blackboard allows ideas from different groups to be written up and compared simultaneously. Within the groups, an effective way to encourage sharing is for the children to work together on one large sheet of paper.

In Chapter 4, some examples are given of this type of organisation with upper juniors.

INDICATING THE TEACHING STYLE IN USE

Despite the growing realisation of the importance of talk in mathematics learning, teachers need not discard altogether their previous classroom techniques, such as exposition and questioning designed to lead children along a particular line of thought. Traditional forms of classroom talk still have a place in mathematics, and they will continue in use for some purposes.

However, because the teacher is beginning to use a variety of teaching styles, it may be necessary to help children to recognise what is expected of them at a particular time – to recognise when a question is a 'checking up' question, which has only one correct answer, and when it is an 'exploratory' question, to which any answer can be built upon in conversation. Some way of conveying the style which the teacher intends to use is necessary, so that the children can recognise what is expected. The seating may be rearranged or the teacher may use subtle signals like standing or sitting to convey different styles. For example, the teacher's introduction may convey messages about how the children are expected to respond. In discussion, the teacher should avoid giving the impression that the guessing game of 'Can you say the word I'm thinking of?' is to be played. Here is a teacher talking with some young children:

> Look, children, I'm holding in my hands a box and it has six flat faces and each of them is a square, and we call this shape a . . . Who can tell me what it is?

This immediately puts the children into a 'right or wrong' situation, where they are invited to guess the name of the shape, to which there is only one right answer. A child either knows the word or does not; there is no room for the negotiation of meaning. A child who already knows the word will have the satisfaction of popping it into the sentence, while those who do not know it have no opportunity to learn by experimenting. The conversation could have developed differently if the teacher had introduced it in this way:

> *Look carefully at the box that I'm holding in my hand. I'll turn it round so that you can see every side of it. Now I'll pass it round so that you can feel it as well – and then I want you to tell me anything you like about the box.*

In this case the children are presented with an open-ended situation, the approach is much more fluid and the talk may continue in any direction. When a child is asked to say something about the box, one possible contribution may be:

> *One of my birthday presents came in a box like that.*

The child's contribution can be gladly accepted without any loss of opportunity to talk mathematics:

> *What was in the parcel, Christopher?*
>
> *A book.*
>
> *Go over to the book corner and see if you can find a book which you think will fit into this box.*

The conversation has been brought to the point where the child's attention is focused on the hollow space inside the box. He will now learn much more about the properties of shape in the discussion which follows than he would have by trying to put one technically correct word in a sentence; the other children will also realise that they are free to contribute to mathematics talk.

SOME WAYS OF INITIATING MATHEMATICS TALK

The teacher's evaluation, with praise for a right answer or disapproval for a mistake, inhibits children from making suggestions or commenting on each other's ideas. An effective way of breaking out of this pattern is to omit the evaluation stage completely. Then the pupils' responses are not inhibited by what they feel to be the teacher's judgments, and the atmosphere encourages tentative offerings from the children. They grow in confidence as they see that their contributions are not evaluated but valued.

Since this style is difficult to acquire, the teacher may need to practise listening skills, alongside the traditional skills of questioning. Some of the following phrases may be found useful in continuing the discussion without resorting to evaluation:

Go on . . .

What made you think of that?

Tell John how it works.

Show me!

I hadn't thought of that.

Do you agree?

What do you think, Saleem?

How did you work it out?

Where did the 8 come from?

The following traditional phrases, on the other hand, can stifle exploration and negotiation of meaning:

That's right.

Good boy!

You're not quite right there.

This is the way to do it.

For the children to benefit fully the teacher has a strong social role to play in managing the discussion: ensuring fair turns at talking, encouraging the shy, helping the diffident to give voice to their opinions, giving a focus to the perceptions of the less fluent children. A particular child who is unsure or reticent can be encouraged to participate by asking:

What do you think, Caroline?

With encouragement, as the children grow older, they learn to manage discussion for themselves, but until then the teacher can tactfully discourage the confident few from dominating the conversation. In this way, the right atmosphere for discussion can be built up, in which the child's attempts to describe concepts are valued, and in which the teacher creates situations which will provoke the development of mathematical language.

DEVELOPMENT IN TEACHING

As the teacher becomes accustomed to the

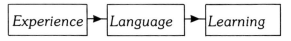

concept of mathematics teaching, this style of teaching will develop further. The teacher can now value children's 'errors' for the insight they give into each child's thinking, and use the errors constructively to point to the next stage in learning.

The value of co-operative group work will also become clearer; children need not necessarily work alone to develop individual understanding. In fact, children can learn a great deal by talking amongst themselves while carrying out mathematical activities, with the teacher joining the discussion at appropriate times.

It is sometimes necessary to move away from the planned direction during a group or class discussion, to follow up an unexpected idea contributed by a child.

Fun with Capacity

Both teacher and pupils need help and encouragement. Unease is often caused by the fact that after a discussion lesson there may be nothing to prove that the children have learned any mathematics. In the absence of written evidence, how will the head teacher, other teachers and the parents know that worthwhile work has taken place?

It takes courage to move away from the safety of the traditional mathematics scheme, but the rewards of mathematics talk are great. Talk gives unequalled opportunities for diagnostic assessment; children can delight the teacher by showing that even at a very young age they are capable of creative mathematical thinking. Conviction will soon grow that purposeful talk plays an important part in the development of thinking skills, and both teacher and children will enjoy mathematics more than had seemed possible. Indeed, the daily written work of mathematics may be a hindrance rather than a help to mathematics learning, as the Thomas Report on ILEA primary schools recommends:

> *Some time now given to writing and to practical work could profitably be given to discussing the implications of practical work, not least in mathematics. The underlying principles in any work should be made as explicit as possible, and this almost always requires discussion.*

Thomas, *Better Schools* (ILEA 1985)

Listening to children, talking with them and often learning with them increases everyone's confidence, insight and enjoyment.

It is difficult to introduce new ideas on one's own, so a teacher who plans to incorporate 'mathematics talk' into the work of the class will find it helpful to co-operate with other interested teachers. Some further suggestions about resources for talk and evaluation are given in Chapter 5.

4 Talking Mathematics in Primary Classrooms

> **Children should be able to make connections between what they already know and new experiences and ideas which they will meet in school. In the classroom, the main vehicle for this will be their own talk.**
> **Children will need help to communicate with and relate to each other effectively.**
>
> *Non-statutory Guidance for English*
> *in the National Curriculum, C3*

Introduction

In this chapter, a number of primary teachers give accounts of ways in which they have used 'mathematics talk' in their teaching. All the teachers have developed their ideas in their own way, the children ranging from those in reception classes to upper juniors.

'Mathematics talk' in a reception class

Teacher and children meet several times a day for class discussion when, in a relaxed atmosphere, they exchange news. Everyone is interested in what each child has to say, and even the shy children can be encouraged to contribute.

Class discussion can be used to introduce activities which lead to 'mathematics talk' so that children can make suggestions about the activity, rather than just listen to the teacher's ideas. In this way they can use their own everyday language.

The composition of groups needs thought; sometimes one talkative child working with others who are quieter may encourage talk within a group. Sometimes it is better to put a group of quieter children together so that each can make a more equal contribution. Occasionally younger children, particularly those who are inclined to work alone, may work well with a leader.

"I have used the following situation to stimulate 'mathematics talk' with my class of thirty five-year-olds: it concerns 3D shapes and their suitability for rolling.

Each child has chosen an item from the classroom which they think will roll. Among the items chosen were a crayon, a bead, a tin, a coin, a pencil, a tin lid, a carton, and a toffee tin and lid.

Many children's choices showed that they knew that the object had to be round to roll. There was a short discussion about shapes and their ability to roll along the floor, using vocabulary such as edge, round, flat, smooth, curved, circle.

Here is an example of one such discussion.

What is wrong with these?

Why aren't they any good for our game?

The teacher shows a box, an animal from the farm and a blackboard rubber.

They will not roll.
That's not round, lumpy, it's got edges, flat, corners.

Children took turns at rolling their objects. There was discussion of the best choices. Why did some turn? Why did some fall over?

The follow-up was done in small groups, away from the teacher so that the conversation was not restricted. The group activities used the same objects but in different situations, so that the language was used more generally. For example, the children took turns to roll their objects down a sloping board.

Here are some examples of the language which occurred:
won't work,
mine went the furthest,
mine went a long way,
that's a good one,
keeps falling over,
it's thin,
different edge,
different direction,
too thin,
it balances.

Things with curves roll

ball

crayon

counter

wheels on the car

round objects roll easily

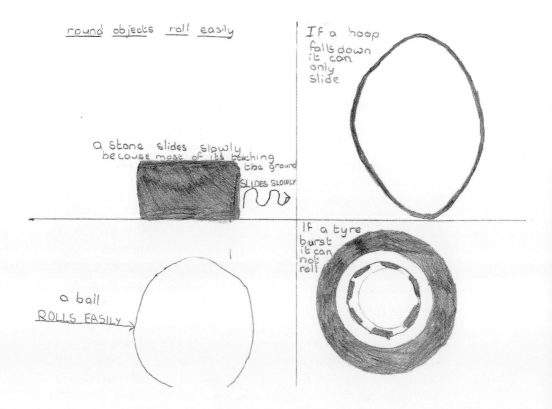

a stone slides slowly because most of its teaching the ground
SLIDES SLOWLY

a ball ROLLS EASILY

If a hoop falls down it can only slide

If a tyre burst it can not roll

Objects easiest to roll:
  beads 'fast',
  tin 'good but turned because of different edges'.

Object hardest to roll:
  coin 'skidded on its side', 'hard to balance', 'thin'.

As I talked to the group, I asked:
Why doesn't a car have wheels which are bead-shaped?

The children's answers to this were:
  too fat,
  they would puncture,
  only a little bit on the road,
  the wheel would bend the exhaust.

Meanwhile, other groups worked at the following activities, which also embodied the idea of rolling:

  Paint a 'silly' car.
  Draw a set of objects which are cylinders, spheres, cubes, cuboids.
  Describe your 'silly' car.

Finally, the groups reported back to the rest of the class, so that there was a further use of the same language.

Children need practice in reporting back, so that they do not feel self-conscious. New words may come from the children, which the teacher has not covered in the planned activity, and more 'childlike' expressions may occur, which are easier for some children to understand."

Toys as a stimulus for mathematics talk

"I am a teacher who had feared and failed at mathematics at school, and I wanted to find a way to teach it so that children would not grow up to feel about it as I had done. Through reading and thinking and through contact with people who enjoyed mathematics I gradually came to see it as an exciting, investigatory, exploratory experience, and I became aware of its pattern, its structure and its creativity.

I felt that young children should approach mathematics through familiar experiences with familiar materials, and that the natural talk that would arise from playing with and handling this material would be conducive to mathematical development, because it would be non-threatening. So I set about making materials for use in my reception classroom: bags of different sizes, shapes, colours and

fabrics, each of which contained a set of soft toys such as caterpillars, beetles, bears, baskets of hens and chicks. Each set had its own attributes of weight, colour, size and number, so that in handling the toys, talking about them, and sorting them, the children's powers of observation, discrimination and logic were developed. Most importantly, I always encouraged the children to talk as they played; the attributes of the toys ensured that much of the talk was mathematical.

The toys were made from many different fabrics, such as velvet, satin, fur, cotton, wool, net and felt, necessitating the use of a wide variety of adjectives when attempting to describe them. The children discovered such words as rough, smooth, scratchy, soft, coarse, harsh, slippery. The main stuffing was foam, but some toys had pebbles hidden inside so that it was impossible to make a judgment about the weight of a toy without picking it up. Small Owl was heavier than his larger friend, while Small Ladybird was lighter than her larger companion, and the family of ladybirds had different numbers of spots. Often, a young child would be startled into comment by holding Large Owl in one hand and Small Owl in the other, and finding that the smaller of the two was much heavier. Adults often say to children 'You are getting big, you are getting so heavy', and so often children start school believing that 'big' and 'heavy' mean the same. The meaning of words has to be negotiated between teacher and child, and the right apparatus encourages discussion about the meaning of words.

Often, when I am talking with the children, I launch into 'once upon a time' language, and at one and the same time I am teaching mathematics and giving the children practice in oral story-telling, an essential ingredient of my reading programme:

> *Once upon a time four creatures who lived in Oaktree Wood met together for a game of Hide and Seek. There was a mouse, a ladybird, a hedgehog and an owl.*

At this point in the story the toy animals are very carefully counted, as there must be no doubt in any of the children's minds that the story is about **four** animals.

> *For a while they played Hide and Seek together quite happily, but then Mouse, who got frightened when he had to close his eyes, refused to take his turn at being the one to seek, and a quarrel broke out. Each of the animals went off on their own to sulk.*

The children's attention is now focused on the fact that 4 can be grouped as 1 and 1 and 1 and 1.

> *After a while Ladybird got very bored on her own and made her way over to Owl. 'Please can I come and talk to you, Owl?'*

Prompted by the question 'What do you say to your friends when you have quarrelled?', the children happily supply the dialogue. I can also help them to focus on the fact that the grouping is now 2 and 1 and 1.

> *On the other side of the clearing Hedgehog went over to Mouse to try to make friends.*

The grouping is now 2 and 2.

> *Alas, Mouse was still feeling very cross and would have nothing to do with Hedgehog, who then made her way forlornly over to Ladybird and Owl to ask if they would be friends with her.*

The grouping has again changed, and is now 3 and 1.

> *The three animals started a competition to see who could tell the funniest joke, and loud laughter came from their side of the clearing. Eventually Mouse could no longer bear to be left out. He*

crept over to his friends to tell them how sorry he was that he had spoilt the game. In return for their forgiveness he promised to behave much better in future.

The grouping is now back to 4 and 0.

Only the skeleton of the story comes from me – all the flesh is put on it by children. As an exercise in group story-telling it describes a situation with which the children are familiar and which can be painful. The accompanying manipulation of the toys shows them very meaningfully how numbers can be partitioned in different ways and then combined again to make the original number. Many other attributes, such as weight, size and position, can be dealt with in similar ways."

Making opportunities for mathematics talk with infants

"In my class of Year 2 children we often spend a few days working on a theme, which may develop from some object a child has brought in, from a casual conversation, or from an event in school. One theme was based on a story about a bald giant and his wig, which I read to the class the day before. I chose the story because it gave a good opportunity to link a 'mathematics talk' activity with the theme.

At the beginning of the day I explained to the class that they would be working in groups, and that they must listen very carefully to the instructions as they would all do each activity at some time during the day. After dealing with questions, I always ask one child in each group to describe to me what the group is expected to do – this can save many problems later.

I gave the 'mathematics talk' group a set of cylinders to talk about and play with, while I checked that all the other groups were settled. The cylinders included a cocoa tin, a kitchen roll centre, a tube which had contained tablets, a (flat) filter paper, a hand cream pot, and a washing-up liquid bottle. We then discussed ways in which the cylinders were the same or different, and when we looked at a very narrow cylinder and a wide one, I introduced the word 'circumference'. This idea was to be a thread which would recur during the term. I had already prepared a card with the word on it, which would later be added to our Mathematics Dictionary.

We used lengths of wool to compare the circumferences of the cylinders, cutting pieces so that they 'just touched' when put round a cylinder. The children protested that wool was not the best thing to use because it stretched 'like elastic', and they decided that strips of paper would be better. I then related the activity to the story, in which a giant was measured for a wig, and the children measured each other for headbands. The headbands were coloured before being put in order of length for mounting on the wall. This 'real' graph might be recalled later when the circumferences of circles were investigated.

While the group were busy, I was able to go round and work with the other groups. Later, the groups rotated, so that all had worked with the cylinders and had added their headbands to the graph before the day was finished. The day ended with discussion of all that had been achieved, the airing of important points that had arisen, such as the size of the giant's head, and the allocation of follow-up tasks.

Later in the term, opportunities were made to find the circumferences of tree trunks, waists, and buckets, and the idea formed the basis for later work with the trundle wheel. The children's repeated use of the word 'circumference' in conversation seemed to indicate that they had become confident of its meaning."

'Mathematics talk' from investigations

"As an advisory teacher, I spend much of my time in classrooms, working alongside the teacher and introducing children and teacher to new mathematical experiences. I gave the following investigation to a class of 10 and 11-year-olds:

> *I bought a carton of orange juice and drank half of it on the same day. The next day I drank half of what was left. On the third day I drank half of what was left again. How long did it take to finish the juice, and how much did I drink each day?*

The children set about working on it in groups of three, and they were asked to put any recordings that they might make in writing or drawing on to poster-sized sheets of paper so that they could be shown to the rest of the class at the end of the session.

The initial reaction was that the juice would last two days as half would be drunk each day. I suggested that the children read the question again more carefully to check that they had fully understood it. Many of the children then drew illustrations to help visualise the gradually diminishing liquid, and one group also found a bottle and used water to do the problem more practically. Soon some children were talking about infinity, and others about the juice never ending. Having established this, they all continued to halve the fractions and to stretch the juice out over more and more days. They were interested to discover the effect on the denominator of the fractions as they continued to halve, and when the numbers became too large for them to handle comfortably they asked for calculators.

One group, rather than using the calculator to double the denominator of the fractions, used it to divide decimal fractions by two, and was surprised that the numbers after the decimal point appeared to be halved, and not doubled as had been the denominators of the vulgar fractions.

At the concluding session, each group in turn reported back, and I drew out the surprising variety of different ideas about fractions and decimals which had arisen during the investigation.

I also suggested to the children that some of them might like to invent their own 'infinity stories', and also to vary the fraction of the remaining juice which was drunk each day. Some children also went on to work out what amount of juice would have been drunk altogether by the end of the sixth or ninth or thirteenth day. It often seems to me that the pooling of ideas, and discussion of contrasting results, is an essential part of the learning experience in 'mathematics talk' sessions.

On another occasion, a class of Year 6 children had been working at intervals on the logical aspects of sets. I reminded them of the work we had done on how to identify a missing attribute block

from a commercially produced set of shapes. They described three or four different strategies for doing this, all of which I accepted as equally valid. I then gave each group of about five children an incomplete set of 'attribute cards', each of which had a face drawn on it, and set them the task of identifying which cards were missing and estimating how many cards there should be in the set; they were asked to draw the missing faces to complete their set. Before they started work I reminded them of the importance of involving all members of the group, and discussing together before launching into the drawings.

There were several more attributes to organise in this problem than there had been with the shape blocks, and each group decided to mount their cards on poster-sized paper in rows and columns for easier sorting. There was a need to talk almost constantly, as each child had to justify why a particular face belonged in a particular position:

That one belongs there because the hair's parted in the middle, the eyes are looking to the right, the nose is filled in, it's smiling and there's no bow tie.

And there were challenges, such as:

> *You can't put that one there because the hair's on the wrong side and the nose is white.*

When one group had exhausted the possibilities and had decided that there were no more faces to be found, I drew the class together to share, to show and to discuss. When it was discovered that each group had different results, there was avid discussion about quicker ways of finding all the faces. Nobody thought they had found all the possibilities and a group volunteered to continue the investigation and report back to the rest of the class at a later date, while another group said they would continue, but use words instead of drawings. The discussion between groups had sparked off further interest in the problem, and it would be possible to explore it again later, at a higher level."

Ways of working

The four teachers who have contributed these accounts of their work are involved with children of different age groups in very different circumstances; two work in deprived inner-city areas with many children whose first language is not English, while the other two work in more affluent suburban areas. However, it is not obvious from their accounts which is which, and certainly 'mathematics talk' is a fruitful way of learning mathematics in very different circumstances.

In spite of the varied situations, the accounts have some features in common. All the teachers value the stimulus of a whole-class discussion to start things off, and to set up an activity which will be suitable for group work, and which will excite the children's curiosity and imagination, thus stimulating a flow of talk as the children work. They all recognise that children have to learn how to discuss mathematics together as they explore a problem, and they provide opportunities for the children to negotiate meaning with the teacher and with their peers. They also value the final 'reporting back' session in which ideas from different groups are pooled, and inconsistencies and arguments can stimulate further investigation. Even a reception class teacher finds that her children can learn to report back from their group without embarrassment, and can express their mathematical thinking with growing clarity and confidence.

TALKING TURTLE

Some computer software available to primary schools encourages mathematical exploration and problem-solving. If children work together on programs of this type, fruitful 'mathematics talk' can ensue, without the need for the teacher to supervise constantly, although it is important that the teacher should join in from time to time, to draw out what has been learned, and to encourage new insights.

A particularly useful example of such a piece of software is LOGO; for young children, a Floor Turtle is a particular stimulus to planning and discussion. In the example given here no teacher was present. The two children worked together in the corridor outside the classroom. Only a video camera was there to record their activities.

Gordan and Steven were not a very well-matched pair of children, although both were six years old and very articulate. Gordan wanted to draw a picture of Concorde starting with the Turtle in the middle of a clean piece of paper. Gordan measured carefully with his hands a distance which he estimated to be 100 Turtle steps, while Steven operated the keyboard.

G   *Forward . . . about . . . forwards what? 200 . . . What d'you think, Steven?*

100 would be about that long . . . 200 . . . 'd be about to there . . . we don't want it there, do we, because the point's got to be quite sharp.

S   *What shall we have then? Just 100?*

G   *No – if we could move it up a bit it would be better.*

S   *Make it go 23 or something.*

G   (In tones of astonishment and scorn) *23!!??? Forward 100.*

S   *All right, then – forward 100.*

Steven was still at the stage of trying numbers at random to see what happened, and it seems likely that he did not yet have a firm grasp of the ordering of numbers in increasing size. Gordan, however, thought ahead, planned and estimated. It might well be that the devastating impact of Gordan's scorn might help Steven to progress. On the other hand, perhaps the time had come to partner Steven with another child, who might be at a similar stage of experimentation to his own, so that conversation could be a more equal exchange. Steven's talk and actions show clearly that he was only able to act as keyboard operator to Gordan, and was not privy to Gordan's careful planning of his drawing. This particular pair of children were not well-matched enough to engage in fruitful 'mathematics talk', although Gordan seems to have started out with good intentions of sharing and consulting fully with Steven.

ONE-TO-ONE 'MATHEMATICS TALK'

In this description of a conversation, a visitor called on a ten-year-old child who was temporarily withdrawn from school; she had been given some mathematics to do so that she would not get too far behind. The visitor's account of the conversation follows:

"Jane said, 'I'm stuck with this. Can you help me?' She was doing an area question which required her to multiply 5 by 3.42. Thinking to help her to a 'quick way' to multiply by 5, I asked her, 'How do you multiply by 10?' She replied, 'Add a nought'. When I asked, 'But what about 3.42×10?' her reply was silence followed by, 'I don't know' after an initial offering of 3.420.

Jane had a calculator beside her so I suggested we look at it and we put 7 into it. I put my finger above the 7 on the display panel while she pressed:

$$\boxed{\times} \quad \boxed{10} \quad \boxed{=}$$

The display showed 70 with the zero below my finger. We repeated this up to 7000 and she got quite excited, saying 'The 7 is moving to the left to make room for the zero.' At 7000 I suggested that we start to divide by 10, and she volunteered that the 7 would move the other way. All went well until we got to 7 ÷ 10. I tried to get her to forecast without much success and the calculator let me down by displaying 0.7 with the 7 still below my finger.

My suggestion that the calculator had shifted everything along so that the figures didn't go off the right-hand end of the display was not too successful.

I wrote down:

$$7 \times 10$$
$$70 \times 10$$
$$700 \times 10$$
$$7000 \div 10$$
$$700 \div 10$$
$$70 \div 10$$
$$7 \div 10$$

She liked the pattern it made and we were able to agree that the 7 would continue to move to the right to give 0.7. She immediately supplied 0.07 for 0.7 ÷ 10 followed by 0.007 etc. and as I had removed my finger from the display panel she now accepted what she saw there.

Now Jane suddenly recollected her 'sums' and said, 'What's all this got to do with my homework?' We returned to 3.42 and successfully multiplied it by 10, 100, 1000 and then divided it by 10, 100, and 1000, getting as far as 0.003 42. When she grasped that she could use this to find 5 × 3.42 without using the calculator, she was quite excited.

I felt at the end of the conversation that the calculator had helped Jane to gain an insight into an aspect of place value that her earlier experience had not given her, and, moreover, that in using the calculator she herself had formulated and verbalised a rule for multiplication and division by 10:

When you multiply by 10 the figures move to the left.

When you divide by 10 the figures move to the right.

Work with a calculator can often encourage the kind of talk that aids understanding of mathematical principles and concepts, provided that there is opportunity to talk with peers and with the teacher while working with a calculator."

CALCULATOR AWARE NUMBER

The CAN (Calculator Aware Number) component of the PrIME (Primary Initiatives in Mathematics Education) project has proved to be a vehicle for encouraging talk in the mathematics classroom. It has been found that, in using a calculator freely, children engage in spontaneous talk with their peers and with their teacher, and that this has provided teachers with an invaluable aid to understanding their children's thinking about number.

Another feature of the project, which relates to this book, is that teachers involved in it have used many of the approaches and ideas advocated here: they have become observers, listeners and facilitators, enabling their children to grow mathematically, whilst at the same time learning more about conceptual stages the children have reached. Classroom organisation within the CAN project resembles much that is described in this booklet to help teachers talk with and listen to their children.

Resources for talking: activities and organisation

5

Opportunities for 'mathematics talk' can arise in a range of classroom settings. With experience we can develop an ability to recognise the mathematical potential of talking through activities in the classroom.

Activities specific to the mathematics curriculum

All aspects of the mathematics curriculum can be used to develop talk which fosters mathematical skills and understanding. The following suggestions are listed under specific topic headings and are starters or generators. Some are genuinely talk activities, others require talk to precede or follow the activity. Once teachers become accustomed to the way in which talk can be generated in mathematical topics, they will find many appropriate examples from their own classroom experience. The activities below are listed under attainment targets of the National Curriculum, but clearly other attainment targets will frequently be involved.

NUMBER AND ALGEBRA (AT2 and AT3)

1   Name some numbers which can be divided by 7 without remainders. As answers are given, use these answers to find new ones, e.g. 700 gives 707 and 693.

2   Phil went to buy three house numbers: 2, 4 and 9. What might have been the number of his house?

3   What numbers can I make using just two threes and any operation?

4   Tell me a number between 14 and 16. Are there any more?

5   What is the smallest number you can think of that is more than 100?

6 What is the biggest number you can think of that is less than 100?

7 Think of a number that leaves 1 when divided by 9.

8 Using Unifix cubes, one per child, find out the biggest class in the school.

9 Have we 1000 books in the library?

10 What could we collect 100 (1000) of?

11 Guess the number I've just picked from the number pack – 'Yes' or 'No' answers only.

12 Your calculator has broken down and only these keys are still working: 5, 2, −, =. How can you use these buttons to make your calculator display each number from 0 to 9?

13 Make up a story to illustrate the number sentence that can be made from $57 - 28$.

14 Tell me anything you can think of about the following number. (Choose any number.)

15 Find as many pairs of numbers as you can which add up to ten and then put them in order.

16 Discuss how to use a simple INPUT/OUTPUT machine.

17 Start with 24 and halve it. Halve it again, and again as many times as you can.

18 Create a pattern beginning 21, 19, 17, . . .

MEASURES (AT2 and AT4)

Time

1 How many hours have you been alive?

2 How many times does your heart beat in a day?

3 Design a calendar that makes your birthday fall on the same day of the week each year.

4 When might my birthday be if it's on the third Tuesday of a month and the date is an even number?

5 Describe a timer you have made which, for example, could be used to time an egg boiling.

6 My daughter is at Junior School and I am three times as old as she is. How old is she? How old might I be?

7 How long would it take you to run a kilometre? How long would it take a hare to run a kilometre?

8 Using copies of the *Radio Times* and the *TV Times* plan your viewing between six o'clock and eight o'clock each evening of the week. Then plan Tuesday night's viewing, allowing time for tea (not in front of the TV), an hour's visit to Grandma's (who has no TV), if you must be in bed by 9.30.

Weight

1 If you lined up counters for 1 kilometre how much would they weigh?

2 How much are you worth? Find your weight in 5 p coins and in 1 p coins.
How much would you rather have?

3 What do 'heavy' and 'light' mean?

4 How would you weigh a goldfish?

5 Find the weights of full packets and tins of food. Do the weights correspond with the marked weights? If not, why not?

6 Which is heavier – a kilogram of bricks or a kilogram of paper?

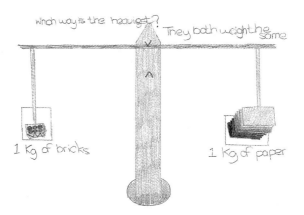

7 I am exactly 20 times as heavy as my cat. How heavy might I be?

8 Using a balance, arrange six parcels in order of weight with the fewest weighings.

9 You need eight blocks and a balance. The blocks are exactly the same shape, size, colour and appearance, e.g. cocoa tins, margarine cartons, etc. One block, however, is heavier than the other seven. You must find that block but only use the balance twice.

10 Investigate the value for money of soap powders.

11 Which five different weights enable you to weigh objects up to 31 grams (whole numbers of grams only)?

12 How much does a Smartie weigh? What weight of Smarties would you need to buy to give everyone in your school one (or ten) Smarties?

Length and Area

1 I need a box with two measurements of 6 cm and 15 cm. Which of these boxes might do? (You will need a collection of boxes.)

2 Would a dinosaur fit into the school hall/playground?

3 What would you use to measure these things: the classroom floor, a stick insect, the distance around your head?

4   How high would a pile of 1000 sheets of paper be?

5   How far would a tortoise crawl in one hour if it didn't stop?

6   A farmer has 40 metres of fencing. Investigate the different shaped fields he could make. Which would give sheep the most grass to eat, assuming the grass is of uniform length?

7   A carpet is 12 square metres. What shape might it be?

8   Can you find out roughly how long a roll of paper is without unrolling it?

9   Find out which of two pieces of paper is bigger. How could you do it?

10   What relationships can you see between the pieces of a tangram?

11   What area of skin do you have?

12   Who can draw the longest line? Measure them and see.

Capacity and Volume

1   Using a set of identical empty washing-up liquid bottles and a litre measure, what fractions of a litre could you make measures for? Which would be impossible?

2   Find the capacity of a sieve.

3   In how many different ways can you find out which of two containers holds more?

4   Find out which of two different brands of shampoo is the better value for money.

5   Find out who has the biggest mouth.

6   Using two sheets of A4 card make two different cylinders with the same surface area. Would each hold the same amount? Make other containers with the same surface area as the cylinders. Compare their capacities. Can you make any generalisations about the relationship between surface area and capacity? What differences are there between making open and closed containers?

7   A box has a volume of 64 cubic centimetres. Use centimetre cubes to show what shape it might be. Make card models of:
    a   a box with the smallest surface area;
    b   a box with the greatest surface area.

8 Find the volume of a stone.

9 Find out how much space each person in this room has to themself.

10 Construct a cone, a cylinder and a cube. Find $\frac{1}{2}$ and $\frac{1}{4}$ of the volume of each of your models.

11 Choose a solid shape and make a box into which it will fit exactly. What can you say about the box and the shape when you have finished?

Money

1 There are five coins in my purse and they total 75p. What might the coins be?

2 It costs £1.20 to see a film, 85p for a cheeseburger, 25p for a can of Coke, 35p for a packet of popcorn, 40p for a choc-ice, 30p for a ballpoint pen, and 15p for a toffee bar. What combinations of these things can I spend my money on if I have £1.75 to spend?

3 How many different ways can you think of to make 83p?

4 In a shop I gave the shopkeeper a £1 coin and he gave me three coins change. How much might I have spent?

5 My book costs three times as much as a newspaper. What is the cheapest price my book could have been? What is the most it could have been?

SHAPE AND SPACE (AT4)

1 Describe how to get to the headteacher's room.

2 Describe a shape (2D or 3D).

3 Explain why a shape rolls or doesn't roll.

4 Describe to a partner a design for them to draw.

5 Using two cuts on a square of paper, how many different shapes can you make? Tell me about them.

6 In how many different ways can you arrange six squares with whole sides touching?

7 Fold a piece of paper (not necessarily in two equal parts) and predict what shapes will be made when you unfold it. Try with two folds. Try with different shaped pieces of paper.

8 What different shapes can you make which have four sides?
Use Geo-strips, strips of paper or card, straws, Geo-boards or drawing.

9 How many 3-, 4-, 5-sided shapes can you make on a Geo-board?

10 How many separate fields can you make using seven pieces of straight fencing? Try with different numbers of pieces of fencing.

11 How many different solid shapes can you make with four cubes?
Try with other numbers of cubes.

HANDLING DATA (AT5)

1 What is the same about all these objects/words/numbers/books/children?

2 Do all wooden things float? How could you find out?

3 We usually line up in two lines (girls/boys). What alternatives are there for making two lines?

4 How can we sort the library books?

5   How many teeth have children in the class had filled? Discuss the findings and ways of recording them.

6   Talk about and interpret a graph constructed by another class.

7   Discuss the best way of recording children's heights so that the younger children in another class can understand the result.

8   Talk about which events are certain and which are impossible. Think of words which describe events lying between certainty and impossibility.

9   If I throw two dice at the same time and add up the results, what might happen?

10 Do you think it will rain tomorrow? What must you think about before you guess?

Mathematics is a powerful tool with great relevance to the real world. For this to be appreciated by pupils they must have direct experience of using mathematics in a wide range of contexts throughout the curriculum.

*Non-statutory Guidance for Mathematics
in the National Curriculum, F1*

Activities which can generate mathematical talk

A wealth of suggestions for cross-curricular activities can be found in the published National Curriculum for various subjects including science and technology, as well as the programmes of study and non-statutory guidance. Opportunities for 'mathematics talk' may be found in other areas of the curriculum.

The following list of suggestions based on infant activities may also be used with older primary children:

- Cooking offers opportunities to discuss weight, time, volume, capacity, besides giving practice in counting, matching, fractions, ratio and proportion.

- Constructional toys and equipment encourage talk about counting and dimensions, shape, area, volume and perimeter.

- Daily or weekly routines of registers, dinner money, banking, tidying up, and the sorting and storing of equipment are rich sources of mathematical activity and talk.

- The PE lesson can encourage spatial awareness through talk about shape, symmetry, direction, speed, height, formation patterns and ordering.

- Art and craft involve two- and three-dimensional shapes, symmetry, scale, tessellations and tangrams as well as talk about quantities for the mixing of paint, etc.

- Music is rich in patterns associated with time and rhythm.

- Displays in and around the classroom can be useful starting points for discussion about dimensions and using space, as well as encouraging interchange of talk amongst children about their own and others' work.

- Planning and preparing for out-of-school visits can include mathematical talk associated with timetables, classification and ordering; the visits themselves offer abundant practical examples of mathematics for talk both during and after the event.

- Discussions about how the children get to school involve choice of route, means of transport and estimation of time of departure to ensure arrival on time.

- The solving of classroom problems or difficulties, such as what to do with the class pet or plants during the holidays, can open up talk which develops some of the strategies and skills of problem-solving so important for mathematical development.

**In life, experiences do not come
in separate packages with subject labels.
As we explore the world around us and live
our day-to-day lives, mathematical experiences
present themselves alongside others.**

*Non-statutory Guidance for Mathematics
in the National Curriculum, F1*

Mathematics in the environment

The activities so far listed, with the exception of visits, have been confined to school. However, the whole environment is rich in potential for mathematical talk!

Two important environmental resources which can be used for generating mathematics talk are:

- the physical environment;
- games and sports.

THE PHYSICAL ENVIRONMENT

The physical environment teems with resources for 'mathematics talk'. Here are a few examples: car wheel hubs (symmetry), manholes (shape and symmetry), patios and pedestrian precincts (tessellations), flowers, trees and birds (classification, area, volume, symmetry, time), houses, buildings and bridges (shape, area, volume, symmetry, rigidity, tessellations), the sea and seashore (time, speed, classification, number, height and elevation, symmetry).

GAMES AND SPORTS

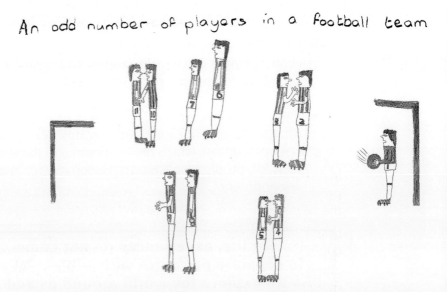

An odd number of players in a football team

Enthusiasts of sports or games will be aware of their potential for mathematics: darts with quick subtractions and doubling and trebling; snooker with its angles and reflections, triangular numbers, computation and estimations; chess with its thinking strategies;

cricket with its analysis of bowling and batting and the spatial awareness in placing the field; the mathematics of planning a knock-out tournament in any game with any number of initial entries; athletics, etc. All these give some idea of the enormous potential involved.

Some of these games are played in school and are used as talk generators, while others can only be talked about. Here is one example of the way in which a game played outside school developed number facility where traditional 'sums' had failed. A young man attending a literacy centre had also expressed an interest in arithmetic. It turned out that he had been tested at a training centre and found to be inadequate in all the arithmetical operations except addition. Given a subtraction to do on paper he used the 'patter' referred to in Chapter 2 and failed to get the correct answer. Subsequent conversation uncovered a splendid competence in subtracting three-digit numbers mentally. He played darts!

Classroom games as a resource for talk

The story above is significant for two reasons: first it indicates the way that talk between teacher and pupil can give the teacher insight into the pupil's thinking; secondly it suggests that games in the classroom might also be used as a classroom resource. At least three types of games may be considered:

- Traditional games which can either be played as they are or modified to become suitable for mathematics.

- Games specifically designed, commercially or by the teacher, to develop a concept or teach a skill.

- Structured play with equipment.

In order to use games successfully in the classroom their mathematical potential must be considered. Overleaf is one teacher's analysis of the part games can play in mathematics.

In playing games, children encounter two aspects of mathematics talk: firstly, the thinking necessary for developing strategies, as in chess, draughts, function-bingo, etc.; secondly, the discussion of moves and strategies.

In order to foster talk, teachers, having introduced a game to a group of children, then ask members of the group to teach others, thus encouraging discussion-skills and questioning. The rules of a game should be flexible enough to allow for adaptations by the children,

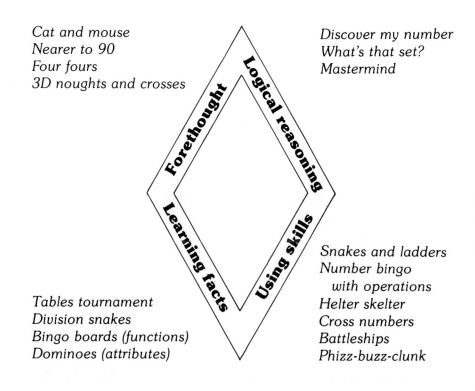

Cat and mouse
Nearer to 90
Four fours
3D noughts and crosses

Discover my number
What's that set?
Mastermind

Forethought
Logical reasoning
Learning facts
Using skills

Snakes and ladders
Number bingo
  with operations
Helter skelter
Cross numbers
Battleships
Phizz-buzz-clunk

Tables tournament
Division snakes
Bingo boards (functions)
Dominoes (attributes)

We talk and play a number game.

THE NUMBER GAME

though participants must understand the rules before the game begins. If the game is a competitive one, pairs or groups of children must be carefully matched to ensure that the same children do not always win. As children grow in competence they may even be encouraged to devise games for themselves, and here earlier practice in adapting rules and teaching each other games will be helpful.

Games and play activities can be devised using various types of structural apparatus available commercially, to generate talk amongst children, both in spontaneous and structured play, thus helping to ensure understanding of the mathematical concepts for which the material is intended.

Making time and evaluating the quality of discussion

6

For some teachers the introduction of strategies designed to make mathematics lessons more strongly language-based will involve quite radical departures from their normal teaching style. These teachers will be changing from a familiar routine in which children practise the four basic arithmetical processes – easy to test by paper and pencil methods – to one in which talk takes on a predominant role in the lesson. This may not in itself be unfamiliar to teachers through their language development work, but it may be new in mathematics.

When embarking on any new approach to teaching it is natural to feel some misgivings both about how to become competent in a new skill and how to evaluate it. Classroom speech provides no written record for teachers or children, for parents or others seeking information about children's progress.

How can we find and use time for talk in our mathematics teaching? How can we evaluate both our efforts and our pupils' progress?

There is no doubt that the arrival of calculators must affect the primary mathematics curriculum. Much primary school mathematics time can be saved by cutting down on written arithmetical processes, thus creating time for newer skills and techniques, including 'mathematics talk', to be learned.

A well-organised classroom in which children are encouraged to be autonomous frees the teacher to spend time listening and talking to children about their mathematics.

Here are two suggestions about organisation made by teachers experienced in language-based learning: one, a chart using a series of questions to help teachers assess their progress in the new teaching style, the other, a list of practical suggestions for use in the classroom.

Some self-help questions for organising talking-time

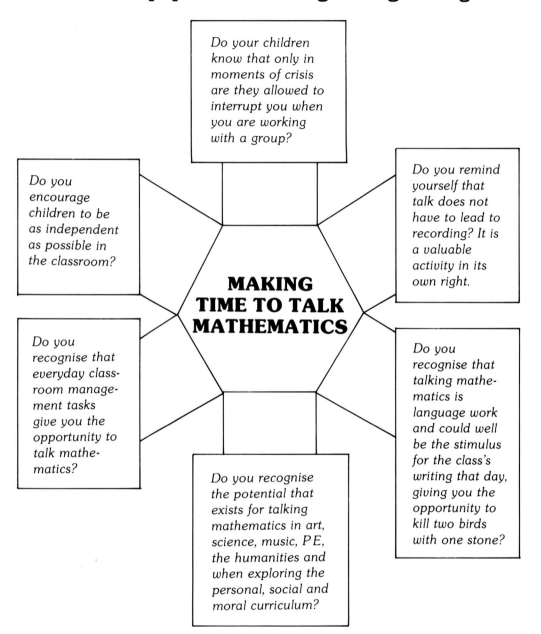

Do your children know that only in moments of crisis are they allowed to interrupt you when you are working with a group?

Do you encourage children to be as independent as possible in the classroom?

Do you remind yourself that talk does not have to lead to recording? It is a valuable activity in its own right.

MAKING TIME TO TALK MATHEMATICS

Do you recognise that everyday classroom management tasks give you the opportunity to talk mathematics?

Do you recognise that talking mathematics is language work and could well be the stimulus for the class's writing that day, giving you the opportunity to kill two birds with one stone?

Do you recognise the potential that exists for talking mathematics in art, science, music, P E, the humanities and when exploring the personal, social and moral curriculum?

Hints for organising talking-time

1 Always have a plentiful supply of paper in different colours, shapes and sizes to which the children have free access.

2 Have tools such as pencils, crayons, glue, paints available and accessible to the children without their having to ask you for them.

3 Have other activities available for children to turn to when they have finished a piece of work and you are still talking with a group.

4 Encourage extending organisational tasks into mathematics activity:

    a after washing brushes they can be sorted into sets, thin/thick paint brushes/glue brushes, before being put away into their containers;

    b registers and dinners – talk about and compare numbers of children absent/present, going home/having a packed lunch/school dinner.

5 Encourage children to tell you about a task they have completed.

6 Involve children in judgments about organisation (it can help their concept of time too) by discussing with them how long it will take to clear up after one activity before preparing for another, e.g. clearing away glue, paint, etc. before preparing to watch a TV programme.

And **above all:**

7 Make sure that the children understand that they must not interrupt talking work except in a crisis. Have firm rules on this.

Evaluating the discussion and its outcomes

Some thoughts should be given to the nature and quality of the discussion and to its further development. The following questions for teachers to ask themselves indicate ways forward:

- How readily is the problem/situation/activity understood?
- Does one child always take the lead?
- What does each child say – issuing instruction, thinking aloud, questioning, etc?
- Do the children listen to each other?
- What questions do they ask me?
- Have I spotted any misunderstandings?
- What about my input into the discussion?
- What mathematical vocabulary is used confidently?

- Are there examples of everyday language being used instead of mathematical language? What does this tell me about their stage of development?

- What am I learning about each child's understanding of mathematics?

- How can I build on this information and how can I use it in my teaching?

- Have I found out anything about a particular child's understanding or knowledge that I didn't know before?

- Did the children enjoy working in this way? What problems did they have and how did they overcome them?

Using tape recorders as a tool for talk and evaluation

In a busy classroom, especially one in which activities leading to language are in use, the teacher's role is a complex one. Since it is impossible for the teacher to be with every group of children in the classroom at the same time, it is important to find organisational devices for getting feedback from other children elsewhere while the teacher is with one particular group.

One way to do this, described earlier in Chapter 4, is to organise the lesson so that class and group discussion alternate, the children taking turns to speak for their group while the teacher moves round the class. Another possibility is to make substantial use of the tape recorder in the classroom for diagnosis and evaluation.

A piece of recorded discussion can provide the teacher with:

- a means of assessing the language skills of pupils;

- a record of work done on a specific topic or activity to measure achievement for assessment purposes;

- an insight into difficulties particular children may be having and into the teacher's own contribution;

- an insight into individual children's level of mathematical understanding and their development of language skills;

- an insight into the effectiveness of the teacher's own contribution.

Skills in the use of a tape recorder

There are, however, skills to be learned in the use of the tape recorder. Children need to be able to record a discussion which can be retained and played back later by the teacher for diagnosis. With practice, the children can learn to plan their discussion to demonstrate what their activity has achieved.

One teacher, eager to tape her children's discussion discovered that, at first, they tended to 'show off' and declaim instead of talking naturally. Nevertheless the resulting recording did indicate interesting language used by the children, and showed the potential of the tape recorder for assisting learning through discussion.

Here are some hints for successful use of a tape recorder which arose from an initial attempt to record some children playing a place-value game:

- make sure the children are relaxed enough to act naturally with a tape recorder present;

- try to encourage them to speak clearly;

- cover tables and surfaces to be used with a cloth to deaden the sounds made by the equipment;

- encourage children to describe moves they make in, for example, playing games, so that they practise the use of mathematical language, thereby helping the recording to make sense to the listener who hasn't got the visual information available to an observer;

- start by having only two children working together at the same task.

It may be necessary at first to isolate the group of children being recorded in order to minimise the extraneous noises that inevitably occur in a busy classroom. Later, it might be possible also to capture the exchanges that come from other children drawn spontaneously to join the group being recorded.

7 Conclusions and the way ahead

Of all significant features of effective learning in mathematics the way pupils identified with the work they were given was of paramount importance. Distinctive, good work in mathematics was generally accompanied by a high level of motivation and engagement in the task: the pupils showed interest, commitment and persistence. This was often reflected in the children's eagerness to talk about the work in hand. In such situations there was a strong sense of the child being largely in control of the work but with support and advice on hand as necessary.

Aspects of Primary Education: The Teaching
and Learning of Mathematics (HMSO, 1989)

The members of the working group are conscious of the enormous contribution parents can make to the development of children's language, including its use in mathematics. We hope many parents will read this book, and also the book *Sharing Mathematics with Parents*.*

It has not been within our brief to look specifically at children who are learning English as their second language, or at children who are bilingual. We feel, however, that all our suggestions will enhance the learning of mathematics in any language. Many groups are already devising mathematical materials for children learning English as their second language.

As a group we have learned most about 'mathematics talk' from discussing what we want children to learn from talking, from listening to tapes and reading transcripts (and then considering what they tell us about children's thinking and learning) and from our heightened sensitivity to what is happening in the classrooms.

We should like to suggest, therefore, that promoting talk in the classroom on the lines we have indicated is easier if several teachers work at it together. In this way you would be able to share experiences, offering support and encouragement to each other as

*Mathematical Association, *Sharing Mathematics with Parents* (Stanley Thornes, 1987)

we have done. Talk amongst teachers is as important as talk amongst children, and between children and their teachers.

Some of the key issues of which we are now more aware include:

- the vital need to find time to listen properly to children's conversation, in order to help them develop their thinking;

- the difficulty of resisting the temptation to tell children answers and save time, as against asking further questions to draw out ideas from them;

- ensuring that questions are structured so that there is often more than one acceptable answer, greater sophistication being expected from the more able children;

- giving the slower thinkers at least as much opportunity to answer as everyone else in the group;

- taking every child's response seriously, which is as important for the self-respect of the child as it is for our understanding of the thinking behind the response;

- learning to avoid telling children they are wrong – if children give different answers to a question, each child has the chance to reassess their response and change it if necessary, and they sometimes have an answer the teacher has not thought of;

- setting up the kind of mathematical situation that generates discussion leading to real mathematical progress;

- providing practical activities using materials that stimulate children's interest;

- organising the classroom to maximise the time available for mathematical discussion;

- enhancing children's confidence in talking mathematically so that they enjoy their learning and develop a positive attitude to the subject.

Opposite is a valuable extract from *Mathematics from 5 to 16.*

DISCUSSION

There is much to discuss in mathematics: the nature of a problem in order to comprehend what is intended; the relevance of the data; the strategies which might be used to produce solutions; and the concepts which need to be clarified and extended. The correctness of results needs to be discussed; where mistakes have been made these should not be ignored or summarily dismissed as they are often profitable points for discussion if handled sensitively. But useful discussion can also take place between pupils without the involvement of the teacher. This is particularly so when they are co-operating in solving a problem, involved in investigative work, carrying out a statistical survey, doing a practical task which requires more than one pupil to complete it, or working together on a micro-computer. The quality of pupils' mathematical thinking, as well as their ability to express themselves, is considerably enhanced by discussion.

HMI, *Mathematics from 5 to 16:* no 3 in the 'Curriculum Matters' series
(HMSO, 1985)

There is no doubt that the development of skills in talking mathematically proceeds in parallel with the same skills right across the curriculum and it is well to remember that:

Speaking and listening are essential to success in adult life. We all need to be able to express our hopes and fears, and solve our problems by discussing them, as well as listening and responding positively to other people. Adults who can persuade others, defend and argue viewpoints, and listen in a discriminating way have enormous advantages. Such skills are fundamental to the social and personal development of young people. It is psychologically important for them to have opportunities to talk and listen in structured ways. Success improves the child's self-image and increases confidence.

Tarleton, R., *Learning and Talking* (Routledge, 1988)

We hope you will be prompted into developing your own teaching by the ideas set out in this book, and that engaging in more 'mathematics talk' will add to your enjoyment in teaching mathematics.

Bibliography

References

Barnes, D. (1976) *From Communication to Curriculum* Penguin.
DES (1975) *A Language for Life: Report of the Committee of Inquiry* (The Bullock Report) HMSO.
DES (1979) *Mathematics 5–11: a Handbook of Suggestions* (HMI Matters for Discussion (9) HMSO.
DES (1982) *Mathematics Counts: Report of the Committee of Inquiry into the Teaching of Mathematics* (The Cockcroft Report) HMSO.
DES (1985) *Mathematics from 5 to 16* (Curriculum Matters 3: an HMI Series) HMSO.
DES (1989a) *Aspects of Primary Education: The Teaching and Learning of Mathematics* (An HMI Report) HMSO.
DES and Welsh Office (1989a) *Mathematics in the National Curriculum* HMSO.
DES and Welsh Office (1989b) *English in the National Curriculum* HMSO.
NCC (1989a) *Mathematics Non-Statutory Guidance* NCC.
NCC (1989b) *English Non-Statutory Guidance* NCC.
NCC (1991) *National Curriculum Council Consultation Report: Mathematics* NCC.
Thomas, N. (1985) *Better Schools* ILEA.
Walter, M. (1984) *A Second Mirror Book* Scholastic Publications.

Some suggestions for further reading:

Books

Association of Teachers of Mathematics (1983) *Language and Mathematics* ATM.
Association of Teachers of Mathematics (1988) *Reflections on Teacher Intervention* ATM.
Ball, G. (1990) *Talking and Learning* Basil Blackwell.
Brissenden, T. (1988) *Talking and Mathematics* Basil Blackwell.
DES (1989b) *Girls Learning Mathematics* (Education Observed 14: an HMI series) HMSO.
Harvey, R., Kerslake, D., Shuard, H. and Torbe, M. (1985) *Language, Learning and Teaching 6: Mathematics* Ward Lock.
Hughes, M. (1986) *Children and Number* Basil Blackwell.
Mathematical Association (1987) *Sharing Mathematics with Parents* Stanley Thornes.
Mathematical Association (1991) *Develop Your Teaching* Stanley Thornes.
Norman, K. (1990) *Teaching, Talking and Learning in Keystage 1* NAPE and NOP.
Shuard, H. and Rothery, A. (1984) *Children Reading Mathematics* John Murray.
Tarleton, R. (1988) *Learning and Talking* Routledge.
The Royal Society and The Institute of Mathematics and its Applications (1986) *Girls and Mathematics* Royal Society.
Wray, D. (Ed) (1990) *Talking and Listening* Scholastic Publications.

Articles

Bain, R. (1988) 'Let's Talk Maths', *Mathematics in School,* 17(2) pp 36–9.
Duffin, J. (1986) 'Mathematics through Classroom Talk', *Mathematics in School,* 15(2) pp 10–13.
Kerslake, D. (1989) 'Collaborative learning in Mathematics' *Mathematics in School,* 18(1) pp 26–7.
Nugent, W. (1990) 'Tomorrow I am going to turn into a giraffe . . .! Or is it Possible to Discuss Probability with 5-year-olds?', *Mathematics in School,* 19(1) pp 10–13.
Smith, R. (1989) 'What is going on in their heads?' *Mathematics in School,* 18(5) pp 33–5.